PLANTS UNSAFE FOR WINEMAKING

by

T. EDWIN BELT

First Impression: December 1972

ISBN 0 900841 31 1

Printed in Great Britain by:
The Standard Press (Andover) Ltd., South Street, Andover, Hants.
Telephone: 2413

PLANTS UNSAFE FOR WINEMAKING

Including

Native or Naturalized Plants, Shrubs and Trees

which are dangerous, doubtful or
dubious as winemaking ingredients, or
which can be used only within limitations

INTRODUCTION

The purpose of this book is to list certain plants, fruits and flowers which should on no account be used for wine-making because they are highly poisonous, and also to indicate others which may or may not be useable, depending upon the process employed. " Poisonous," after all, is a relative term ; plants which are poisonous to humans may not be poisonous to animals, and vice versa, but in outline context it means plants that have a chemical content beyond that which can be tolerated by the human digestive system.

Many plants and fruits in everyday use, of course, contain poison, or toxic substances, in some degree. There is, poison, for instance, in potatoes if they are " green," in rhubarb leaves, in peach and plum stones, and even in beech nuts and young apple pips.

This, you may say, hardly makes the fruit " poisonous" ; whoever heard of anyone dying through eating apple pips ? Yet there is on record the case of a man who liked them so much that he had saved up a cupful for himself as a special treat. He ate them—and died from prussic acid poisoning, a most painful death. So should we list the apple as dangerous?

One also comes up against the curious fact that many materials that are regularly used for winemaking contain some toxic substance yet wine made from them appears to be safe. The answer may lie in the winemaking process itself ; some poisons may be vitiated by boiling, some counter-acted by the presence of sugar, and others perhaps by the fermentation process itself. Since little or no scientific re-search has previously been done in this field, just how much of the original toxicity of a must " comes over " in a finished wine remains unknown ; I am now doing a series of chemical analyses of certain " unsafe " wines, but it will be a long job !

A further complication is that winemakers often use only small amounts of certain materials, and that the poisonous substance is thus present in only negligible quantities. Many ingredients, it must be pointed out, contain small amounts of substances which in other circumstances and in quantity would be labelled " poisons."

I have consequently thrown my net wide, and wherever danger can conceivably exist have tried to indicate it, choosing the description " unsafe " for winemaking (rather than " poisonous") deliberately. It is better to be safe than sorry, and to that extent " Plants Unsafe for Winemaking " cannot fail to serve a useful purpose.

The final responsibility must rest with the winemaker, of course, because to a large extent whether a wine is safe or not will depend upon how much of the " doubtful " material he uses, *how* he uses it, and how much of the finished wine he drinks. Even then, the effect upon one person may be different from that upon another: the truism "One man's meat is another man's poison" certainly applies here.

Finally, I cannot assert that because a plant is listed here it will certainly produce unsafe wines. By the same token, the reader must on no account assume that because a plant is NOT listed in this book it is safe ; obviously any such list is far from exhaustive.

T. Edwin Belt

ACACIA, FALSE
Robinia Pseudoacacia

Tree reaching a height of 26 metres found in parks. Deeply fissured bark and very delicate foliage. Flowers white or pinkish, and mature into peapod-type fruits a decimetre long. Bark, foliage, shoots and seeds are all poisonous.

ACONITE (Monkshood, Wolfsbane)
Acenitum Hapellus Angelicum

All parts are poisonous, due to the presence of alkaloids, and the poison persists after dehydration of the plant. Found in shady places in the south-west of the British Isles, and there are cultivated varieties. The flowers are blue-lilac coloured, several to a stalk, and five-petalled, one petal being larger than the rest and in a helmet shape. They appear in June. Grows up to 1m. 25cm. (4ft.) tall. The black roots are not to be confused with Horseradish, nor the bright green leaves with Parsley.

ACONITE, WINTER
Eranthis Hyemalis

All parts are poisonous, due to the presence of an alkalod with a burning taste. To be found in woods and plantations throughout the British Isles. The flowers are bright yellow, with six petals, and appear during January, February and March. The plant grows to a height of 15cm. (6in.). The only leaves on the flower stem grow immediately below the flower, and the other leaves appear after the flowers.

ALDER, BUCKTHORN, OR BLACK, OR BLACK DOGWOOD
Frangula Alnus

The berries, the bark, and the leaves, are all poisonous, due to the presence of glycosides, which are violently purgative. This is a common shrub in England, growing up to 3 metres tall on peaty soil. The flowers are a whitish-green, with five petals, which are pointed at the extremity, and appear in May, June and July. The colour of the berries progresses from green to red to black. The leaves are oval-shaped, pointed at the stem and rounded at the extremity, and the berries grow at the stem.

ALFALFA OR LUCERNE
Medicago Sativa
A mildly poisonous plant in all parts. Better to be safe than sorry in this instance. Grown as a fodder crop, and escapes into the hedgerows. The deep purple flowers grow on separate short stalks off a main stalk, in clusters, and blossom from April through to August.

ALMOND, BITTER
Three of these kernels will cause severe poisoning. Contains a resultant prussic acid from the glycoside amygdalin.

ALOCASIA
Alocasia
To be avoided in winemaking, due to the incidence of insoluble calcium oxalate crystals. An ornamental plant, of which there are several varieties. The stems and leaves of the variety Macrorrhiza are eaten as a vegetable after the poison has been boiled out.

AMARYLLIS OR BELLADONNA LILY
Amaryllis Belladonna
Contains toxic alkaloids
A pot plant, of which the scented flowers appear first, to be followed by the leaves. There are a large number of decorative crosses of this plant.

ANEMONE, BLUE
Anemone Apennina
Poisonous due to the presence of proto-anemonin in leaves and sap giving an acrid taste. Occasionally found in woods as a garden escape, the eight-petal flowers appearing in April and May.

ANEMONE, SHAGGY, OR PASQUE FLOWER
Pulsatilla Vulgaris
Poisonous—as blue anemone. Found on chalk sub-soils in south-east England. Grows to a height of 25 cm. (10in.) and the six-petal flowers are violet in colour, with a yellow centre, on thickish stalks, appearing in April and May. The leaves are featherlike.

ANEMONE, WOOD, OR WINDFLOWER
Anemone Nemorosa
Poisonous as blue anemone. Commonly found in woods, but not on acid soils and is cultivated. Grows to a height of 15cm. (6in.) and the solitary terminal six-petal white or lilac flowers appear in March and April.

ANEMONE, YELLOW WOOD
Anemone Ranunculoides
Poisonous—as blue anemone. Found in woods on alkaline sub-soils as a garden escape. Grows to a height of 15cm. (6in.) and the six-petal golden-yellow flowers appear in April and May, singly or in pairs.

APPLE
The pips yield prussic acid, so that it might be thought desirable to remove them before winemaking, when making this wine, or of cider. But it is only fair to say that thousands of gallons of apple wine must have been made without this precaution, and we have heard of no resultant deaths !

APPLE, THORN, OR DEVIL'S, OR JIMSON WEED
Datura Strainonium
The alkaloids in the leaves, capsules and seeds, are highly poisonous, and the poison is resistant to drying. Found on cultivated ground in southern England. Grows to a height of one metre. The long and tubular white or violet flowers appear in July through to September, to be followed by the prickly capsule containing black seeds.

APRICOT
The kernel yields prussic acid, and is best removed prior to winemaking with the flesh.

ARTICHOKE, JERUSALEM
Helianthus Tuberosus
The leaves are suspect, and are doubtful for winemaking. There are a number of related species which are grown as ornamental flowering plants. The edible tubers contain inulin and not starch.

ARUM, BOG
Calla Palustris
Contains irritant crystals of calcium oxalate, and is best avoided. Found in south-east England around ponds and other very damp situations. The leaves are roughly a broad heart shape and the minute flowers are crowded on a spike.

ASARABACCA
Asarum Europaeum
A poisonous member of the Birthwort family. Found in woods. The leaves are roughly semi-circular, and the purple bell-shaped flowers appear from May through to August. Each very short stem carries two leaves and one flower.

ASH TREE
Fraxinus Excelsior
The flowers and fruit are probably poisonous, and to be avoided in winemaking, although fermentation perhaps breaks down the poison content. The fruit is one-winged, and the flowers are without petals, having a crimson-coloured stamen. The black buds are a certain recognition sign. Found in woods on chalky soils, but is often planted.

ASH, MOUNTAIN (OR ROWANBERRY)
Rowanberry wine is certainly made but in my opinion it is best avoided, although fermentation probably breaks down the poison content of the bitter berries, which in any event are used in only small quantities.

ASPHODEL, BOG
Narthecium Ossifragum
The whole of this acrid-tasting plant may be poisonous, and it is to be avoided for winemaking. Commonly found in bogs on acidic moors, mostly in the north-west of Britain. The bright yellow-star-like multi-petalled flowers grow up the apex of a leafless stem, and appear in July to August. The leaves are linear, rigid and pointed.

12

Asphodel, Scottish
Tofieldia Pusilla
May be poisonous, and is to be avoided for winemaking. Found in northern England, usually on the banks of mountain streams. The leaves are narrow and pointed. The clean flower stalk carries greenish-white flowers at the tip, which blossom from June to August.

Azalea, Mountain or Trailing
Loiseleuria Procumbens
Contains a neutral principle, andromedo, which is a toxic poison, and is found in the leaves and flowers. Found on high ground in Scotland. A procumbent shrub with small tough evergreen leaves and small pink, five-petalled flowers, which blossom from May to July.

Baneberry, Common, or Herb Christopher
Actaea Spicata
Poison causes extreme purgation and is found in the berries. The leaves can be confused with those of the Elder. Mostly found in woods, in the limestone areas of Yorkshire, Lancashire and Westmorland. Grows up to 60 cm. high, with terminal groups of minute white flowers which appear in May, to be followed by the green berries which turn a shiny black on ripening.

Baptsia or False Indigo
Contains active poisonous alkaloids in all parts of the plant. The flowers are blue, and appear in sparsely formed spikes. Similar in appearance to the Lupin.

Barberry, Common
Berberis Vulgaris
Doubtful for winemaking. Found in hedges and wooded areas. Grows to a height of 2 metres. The flowers are yellow and grow in loose spikes, appearing in May and June. The red berries are oval-shaped.

Barley
Hordeum Vulgare
Should always be boiled when used for winemaking,

to avoid trouble from nitrites, but any danger existing is more probable in America, and is a remote possibility in Britain.

Bay Tree, Dwarf, or Spurge Olive, or Mezereon
Daphne Mezereum

All parts of this plant secrete a very irritant poison which is not removed by drying and storage. The scarlet-coloured berries are particularly high in poison content and are the more dangerous in that they bear a close resemblance to red currants. This shrub grows up to 60cm. in height, and bears sweetly fragrant purplish-red flowers which appear before the leaves in very early spring. The narrow leaves are up to 7cm. long, and grow at the extremities of the branches. Found in gardens, but occasionally grows wild in southern areas of the British Isles on chalky sub-soils.

Bean, Castor
Ricinus

Contains a poison which is removed by heat. This poison is not soluble in oil ; hence castor oil is quite safe.

Bean, Precatory or Rosary Pea
Abrus

The poisonous constituent is inactivated by heat, but one seed can kill a human being, and the plant is best avoided. The ovoid beans are 10cm. long with a one-third black gloss enamel and two-thirds bright red enamel style of appearance.

Beech
Fagus Sylvatica

The nuts are poisonous. This tree favours calcareous subsoils, and reaches a height of 40 metres. The nuts are triangular and appear out of their woody, four-section outer casing in June—July. The leaves are ovate and silky in texture. The flowers appear in March and April in multi-clusters on separate stalks, and the male flowers are catkin-like.

Beet, Sea
Beta Vulgaris Maritima

Best avoided. Contains soluble salts of oxalic acid which

14

combines with calcium to form the irritant insoluble calcium oxalate crystals. The addition of common chalk in wine-making will obviate the risk of the blood being denuded of calcium, but the irritant crystals remain. The roots may be used if not subjected to the action of hot water in forming nitrates, and are best used after long storage, if at all. Found on the sea-shore. Thinly flowered spikes, blossoming from July to September. The green-coloured leaves are fleshy, with a high gloss.

BEET, SUGAR
Beta Vulgaris
The tops are doubtful, but the roots can be used, preferably after Christmas, when the use of precipitated chalk will not come amiss.

BETONY, WATER, OR FIGWORT, WATER
Scrophularia Auriculata
This family are known to be violent purgatives, and Water Betony may not be an exception. It is best avoided Found near ponds and ditches everywhere. The leaves are rounded and arrowhead in shape, and their margins are serrately rounded. The small red flowers grow in groups, each group with a separate stalk from the main stem.

BINDWEED, COPSE OR COPSE BUCKWHEAT
Polygonum Dumetorum
Contains a sharp acrid juice, which is a severe irritant of the intestines. Found in hedges and thickets in the south of England. The twining stem grows up to 1.5 metre tall, and the flowers appear from July to September. The leaves are spear-head shape.

BINDWEED, BLACK
Polygonum Convolvulus
Contains a sharp acrid juice, which is a severe intestinal irritant. Found on cultivated ground.
The twining stem grows up to one metre long, and the flowers appear from July to September.

BINDWEED, FIELD, OR LESSER
Convolvulus Arvensis
Doubtful for winemaking. A climbing plant found on cultivated soil, except in the north of Scotland. The pink and white bell-like flowers appear from June to August.

BIRD OF PARADISE
Poinciana
Poisonous, particularly the pods. This is a shrub having clusters of crested, large, orange and blue flowers, growing on stems up to 1.7 metres tall. A pot plant grown in heated greenhouses.

BIRTHWORT
Aristolochia Clematitis
Best avoided for winemaking until more is known about it. Not a very common plant, growing up to 60cm. (2ft.) tall. The dull yellow flower tubes appear from June to September, to be followed by the pear-shaped fruit.

BISTORT, OR SNAKEWEED
Polygonum Bistorta
Belongs to a poisonous family, and is best avoided until more is known about it. Found in moist meadows. The pink flowers appear in a terminal dense cylindrical cluster from June to August.

BISTORT, ALPINE
Polygonum Viviparum
Belongs to a poisonous family, and is best avoided. Grows on mountain pastures in northern England and in Scotland. The erect stem grows to a height of 30cm. (1ft.) and the reddish-pink flowers appear in a terminal cylindrical cluster from June to August.

BISTORT, AMPHIBIOUS OR AMPHIBIOUS PERSICARIA
Polygonium Amphibium
Comes from a poisonous family and is best avoided until more is known about it. Found in pools and slow-moving streams. The pink flowers appear in an inverted U-shape spike from July to September.

BITTERSWEET, OR WOODY NIGHTHADE
Solanum Dulcamara
The poisonous principle solanine is present in all parts of the plant, sometimes confused with Deadly Nightshade.

Common, climbing over shrubberies and hedgerows, in copses and marshes, where damp and shade are available, in England and Wales and locally in Scotland. Grows to a height of 4 metres in a shrubby, climbing and trailing form. The small, drooping five-petalled flowers grow in clusters, are bluish-purple in colour with bright orange centre-pieces, and appear from June to August, to be followed by the berries, which are scarlet in colour when ripe.

BLEEDING HEART
Dicentra
Contains poisonous alkaloids. Cultivated.

BLOODROOT
Sanguinaria Canadensis
Ornamental garden plant. Contains poisonous alkaloids. Root used in dried state as an emetic.

BLUEBELL, OR WILD HYACINTH
Endymion non-scriptus
Purgative and irritant poison which may be allayed by the action of sugar, but research needed. Other members of the same family are definitely poisonous, particularly the bulbs. Common in woods and shady places throughout Britain. The well-known blue flowers appear in April, May and June.

BORAGE
Borago Officinalis
Doubtful in quantity for winemaking, although it is used as a salad garnishing, and in claret cup and other drinks in the same manner. Cultivated as a herb and garden flower. The flowers are bright blue with a white ring.

BOX TREE
Buxus Sempervirens
Causes irritation of the digestive system, and is the most

dangerously poisonous of the hedge plants. All parts of the plant contain the poisonous Buxine. Grows to a height of 4.5. metres on chalk subsoils in southern England, and is cultivated elsewhere. Dark, glossy, green, oval, small leaves. The white flowers grow in semi-circular clusters, and appear in April and May.

BRACKEN OR BRAKE
Pteridium Aquilinum

All parts poisonous due to the presence of thiaminase, which is not removed by drying and storage. Grows to a height of 2 metres, and the fronds are well-known, appearing from May to October. Found on heaths and hillsides throughout Britain.

BROCCOLI

If used for winemaking small amounts should preferably be used ; it does not appear to be a promising winemaking material anyway. (See cabbage.)

BROOM, COMMON
Sarothamnus Scoparius

Contains neglible (to the prudent winemaker) quantities of the poisonous alkaloids cytisine and sparteine, and the glycoside scoparin, which are mostly found in the seeds. The flowers, used in moderation, should not cause ill-effects in wine, but there have been reports of cases of sickness after drinking wine made from *dried* broom flowers. An erect shrub, up to 2 metres tall, widely distributed on dry heaths

and also cultivated. The flowers are large, bright yellow, sometimes tinged with red, having five unequal petals, and appear in May and June.

BROOM, SPANISH OR LINK
Spartium Junceum

As Broom, Common (which see). Similar in appearance to Broom, but the stems are round and the leaves thin. Cultivated.

BROOMRAPE, AMETHYST-BLUE
Orobanche Amethystea

A drastic purgative. Found in south-east England, but

18

chiefly in the Channel Islands. The massed flower spike blooms in June and July.

BROOMRAPE, BRANCHED
Orobanche Ramosa
A drastic purgative. Appears in the Channel Islands and in south-east England. The stem has a few basal branches. The dense flower spikes have cream-coloured flowers with purple edges, and the blossoms appear from July to September.

BROOMRAPE, IVY
Orobanche Hederae
A drastic purgative. Found mainly on the coastline of southern England and Wales. The flower petals are cream, veined with purple, and form a spike.

BRUSSELS SPROUTS
Best used in winemaking (if at all, since it is hardly a suitable material) in only small quantities. (See cabbage.)

BRYONY, BLACK OR BLACKEYE ROOT
Tamus Communis
This is a member of the Yam family, and the Red Bryony is a member of a totally different family, the Melons, but both harbour narcotic irritants in the root and berries. Common in woods and hedgerows in the midlands and southern England, and in Wales. The twining stem will grow to a height of 4 metres where it can find support. The small whitish-yellow-green flowers are sparse in spikes, and appear from May to July. The berries turn from green to bright scarlet.

BRYONY, RED, OR WHITE, OR DEVIL'S TURNIP, OR WILD VINE
Bryonia Dioica
This is a member of the Melon family, as distinct from Black Bryony of the Yam family, but nevertheless both harbour narcotic irritants, chiefly found in the roots and berries. Common in the hedgerows of south-east and the midlands of England. Climbs by means of tendrils to a height of 3 metres. The small flowers are greenish white with five petals and bloom from May through to August. The berries are red.

BUCKEYE
Aesculus
Poisonous. Similar to the horse-chestnut.

BUCKTHORN, COMMON, OR PURGING
Rhamnus Catharticus
The berries contain poisonous, purgative, glycosides, and the plant is poisonous in all parts. A shrub growing up to 5 metres tall, found in hedges on calcareous soil in south-east England. The branches are opposite and end in thorns. The flowers are small four-petalled and green, grow in clusters at the base of the leaves, and bloom in May, June and July, to be followed by the green berries which turn black on ripening. The berries are not to be confused with those of the Elder ; they grow from the base of the leaves, unlike those of the Elderberry. The Buckthorn is character-ised by its thorn.

BUCKTHORN, SEA
Hippophae Phomnoides
Best avoided for winemaking until more is known about it. Found on sand dunes on the east and south-east coasts of England. A thorny shrub growing to a height of 2.5 metres. The flowers are small and green, appears in March and April to be followed by the globose orange-coloured fruit.

BUCKWHEAT
Fagopyrum Esculentum
The black grain is harmless, but beware the rest of the plant. However, allergic rashes do occur when buckwheat flour is used. Grows on cultivated ground. Reaches a height of 60cm. (2ft.) with a slender reddish stalk and reddish flowers, which appear in July and August.

BUDDLEIA
Buddleia Davidii and Alternifolia
Poisonous ; contains strychnine, in association with brucine, in the seeds. The flower clusters are lilac coloured, and grow on a bush.

BUGLOSS, SMALL
Lycopsis Arvensis
Best avoided, in common with Viper's and Purple

Viper's Bugloss. The small pale-blue flowers have a white centre. The oblong leaves are bristly. Common in cornfields.

BUGLOSS, VIPERS
Echium Vulgare
Best avoided, as is the purple variety, since they are possibly poisonous in all parts. Common everywhere where the subsoil is chalk. Grows up to one metre tall on a round, hairy stem, up which the bright blue flower grows intermingled with the leaves.

BUGLOSS, VIPER'S PURPLE
Echium Plantagineum
Might cause toxic liver cirrhosis, and is best avoided. The plant bears long, hard bristles throughout. The purple flower has been said to resemble an open mouth.

BUTTERCUP, BULBOUS OR ST. ANTHONY'S TURNIP
Ranunculus Bulbosus
Contains an irritant poison when fresh, but is safe if used after drying. The bulbous roots are equally poisonous. The protoanemonin poison has an acrid, burning taste. Common throughout Britain. Flowers from April to July and is recognised by the bulbous swelling at the base of the stem.

BUTTERCUP, CELERY-LEAVED OR CROWFOOT
Ranunculus Scleratus
Most poisonous of all the buttercups, solely attributable to its luxuriant growth. Contains an irritant poison with an acrid taste when fresh, but is safe if used after drying. Found by the side of water, the flowers attaining a height of up to 60cm. (2ft.) with foliage up to 30cm. (12in.). Flowers appear from May to September.

BUTTERCUP, CREEPING
Ranunculus Repens
Contains an irritant poison when fresh, which is at its maximum at flowering time. Safe after drying. The protoanemonin poison has an acrid, burning taste. Common on British pastures. Reaches a height of 60cm. (2ft.). Flowers

from May to August. The rooting runners are a recognition sign.

BUTTERCUP, HAIRY
Ranunculus Sardous
Poisonous—as buttercup (creeping).
BUTTERCUP, MEADOW OR COMMON UPRIGHT CROWFOOT,

GOLD KNOTS
Ranunculus Acris
Poisonous—as buttercup (creeping). Found everywhere. Reaches a height of one metre. Flowers appear from May to August.

BUTTERWORT
Pinguicula Caudata
Doubtful for winemaking. The large red flowers can be seen in botanical gardens, and are a rare pot plant.

BUTTERWORT, COMMON
Pinguicula Vulgaris
Doubtful for winemaking, and the taste is disagreeable. The flowers are blue, have a long spur, and employ a sticky substance with which to capture insects, which are then digested as a source of nitrogen. There is a variety, P. Lusitanica, having pinkish-lilac and yellow flowers. Both grow on wet heaths and bogs.

CABBAGE
If used for winemaking—and it does not appear to be promising winemaking material—it is preferably used in only small amounts.

CABBAGE, SKUNK
Lysichiton Symplocarpus
Contains irritant crystals of calcium oxalate which have resulted in loss of life due to swellings inside the mouth which inhibited breathing. A marsh plant with very large leaves, and yellow flower spikes.

CABBAGE, WILD, OR KALE
Brassica Oleracea
In common with other members of the cabbage family,

including the cultivated varieties, it is surprising to learn that, if consumed in large quantities, and in conjunction with an iodine-deficient diet, goitre can be the outcome : wild cabbage can give rise to anaemia, and is best avoided. Found on the sea cliffs of south and south-west England, and of Wales. Grows to a height of 60cm. (2ft.). The yellow flowers are four-petalled in a long spike, and bloom from May to August.

CAMAS, DEATH
Zigadenus

Poisonous. Grass-like leaves. The flowers are white, greenish or pink, similar to a very small lily, and grow in spikes. The plant springs from a bulb.

CAMPION, NIGHT-FLOWERING
Silene Noctiflora

Grows to a height of 6 dcm. The flower petals are rosy-coloured on top and yellowish underneath, and open at night. Poisonous.

CAMPION, BLADDER
Silene Vulgaris

Grows to a height of 9 dcm. on a woody stem. White fragrant flowers, oval leaves. Poisonous.

CAMPION, MOSS
Silene Acaulis

The five-petalled purple-pink flowers grow abundantly in a mass of green foliage close to the ground, which arises from short woody stems. Poisonous.

CAMPION, WHITE OR EVENING
Silene Alba

Grows to 1 metre tall on a woody stem. The white flowers are 3cm. across and evening-scented. Poisonous.

CAMPION, RED, OR MORNING
Silene Dioica

Erect flowering stems up to 9 dcm. tall, among creeping shoots. Bright rose-coloured flowers up to 25mm. diam. Scentless. Poisonous.

CANDELABRA CACTUS
Euphorbia
Poisonous. Cultivated.

CARDINAL FLOWER
Lobelia
Poisonous. Cultivated. Carries tall spikes of vivid scarlet flowers, and the foliage is bronze-red. Blossoms from July to October, when it will have attained a height of 60cm. (2ft.).

CARNATION
Dianthus
Contains poisonous saponins, but in amounts not likely to be dangerous in the quantities used for winemaking.

CARROT, WILD
Daucus Carota
Doubtful, and best avoided in winemaking. The whitish flowers appear in May and June in clusters on long, separate stalks. Grows on the coast on chalk subsoils.

CASSAVA
Contains cyanide which is removed by heat. Should be avoided for winemaking. A tuber rather similar to the potato.

CASTOR OIL PLANT, CHEESE OR WEEPING PLANT
Ricinus Communis
The seeds are extremely poisonous, due to the presence of ricin. The plant is found in this country only when required for ornamentation of the garden, and as an indoor potted plant. Grows up to 2.7 metres tall. The leaves have anything from five to 12 lobes and exude drops of liquid. The flowers appear on the apex of a thick stalk in July.

CAULIFLOWER
Best used in amounts not exceeding a normal meal intake. (See cabbage.)

CELANDINE, GREATER, OR POPPY CELANDINE, OR SWALLOW WORT
Chelidonium Majus

Contains poppy alkaloids, which render this plant dangerous for winemaking ; these active principles are very toxic. This perennial plant is common throughout Britain, and favours ancient building sites. Recognisable by the foetid yellow juice in the stems, which turns red when the stem is broken open. Grows to a height of one metre. The yellow flowers have four petals and appear from May through to August.

CELANDINE, LESSER, OR PILEWORT, OR BRIGHTEYE OR GOLDEN STARS
Ranunculus Ficaria

The irritant poison protoanemonin renders this plant unfit for winemaking. Common in moist shady places throughout Britain. Grows to a height of 15cm. The solitary flowers have eight glossy golden-yellow petals and appear from March to May.

CELERY

Grown where a heavy application of nitrate fertilizer has been made can render this plant unsuitable for winemaking if the amount to be consumed as wine is greater than a normal meal intake.

CELERY, WILD
Apium Graveolens

Best avoided in winemaking. The flowers appear in June, July and August ; they are whitish, and grow in clusters. Appears mostly near the sea, and favours wet places.

CHAFFWEED
Anagallis Minima

Doubtful for winemaking. Grows up to 8cm. in height. The flowers are minute and pinkish in colour, appearing in June and July along the delicate stem. Appears on sandy ground, chiefly in the south-west of Britain and the west of Scotland.

CHARLOCK, WHITE, JOINTED, OR WILD RADISH
Raphanus Raphanistrum
Very dubious for winemaking. The flowers are white, pale yellow, or pale lilac, growing on a plant which does not exceed 60cm. (2ft.) in height ; they are four-petalled, the petals widely spaced, and appear from May through to September. Grows on cultivated land.

CHARLOCK, WILD MUSTARD, OR KEDLOCK
Sinapis Arvensis
Poisonous when the seed pods have formed. Bright yellow flowers common in cornfields, at the uncultivated margins of which they blossom throughout the summer, from May to August. Grows to a height of 60cm. (2ft.).

CHERRY
Remove the stones when winemaking and so avoid the resultant prussic acid from a breaking down of the constituent glycosides.

CHERRY, BIRD
Prunus Padus
As for " cherry," q.v.

CHERRY, CORNELIAN
Cornus Mas
See " cherry."

CHERRY DWARF
Prunus Fruticosa
See " Cherry."

CHERRY, JERUSALEM
Solanuam Pseudocapsicum
Poisonous. This is an indoor plant which produces attractive bright orange to red berries in December.

CHERRY, SOUR
Prunus Cerasus
See " Cherry."

CHERRY, WILD (Gean)
Prunus Avium
See " Cherry."

CHERVIL, BUR
Anthriscus Caucalis
A plant to view with suspicion. The minute flowers appear in June and July. Frequently seen in seaside hedges in south-east England.

CHERVIL, ROUGH
Chaerophyllum Temulentum
A plant to view with suspicion. The stem is purple-spotted and hairy. The flowers appear in June and July in sparse clusters. Found near hedges in England.

CHESTNUT, HORSE
Very poisonous.

CHICKWEED, BEARDED MOUSE-EAR
Cerastium Brachypetalum
A member of the poisonous Pink family, and to be viewed with suspicion. The flowers appear in May.

CHICKWEED, COMMON
Stellaria Media
Poisonous. The small whitish flowers appear all the year round. Found on cultivated ground everywhere.

CHICKWEED, DWARF
Moenchia Erecta
Poisonous. Grows up to 10cm. (4in.) high. The pale yelllowish four-petalled (widely spaced) flowers appear from April to June. Grows on sandy or similar places in England.

CHICKWEED, JAGGED
Holosteum Umbellatum
Poisonous. Grows to a height of 15cm. (6in.). The flowers are white or pale pink and appear in April and May. Found growing out of ancient ruins.

CHICKWEED, WATER
Myosoton Aquaticum
Poisonous. The ten-petalled whitish flowers appear from June to August. Grows in wet places.

CHICORY
Cichorium Intybus
Doubtful for winemaking. Grows up to one metre high. The blue flowers appear at the base of the leaves from July to September. Appears in fields and waste ground where the subsoil is chalk.

CHIVE
Allium Schoenoprasum
Large amounts of this plant will cause anaemia, as is the case with the onion. This is rather surprising, but in mitigation it must be said that the resultant strong flavour should inhibit such lavish use.

CHRYSANTHEMUM
Best avoided. Pyrethrum, a potent insecticide, is obtained from the flower heads of a daisy-like flower of this genus.

CHRISTMAS CHERRY
Solanum, Capsicastrum
Contains the poisonous principle solanine. The red berries make this a favourite pot plant.

CINERARIA
Senecio Cineraria
To be avoided by the winemaker. The yellow flowers appear in clusters from June to August. The leaves are felt-like on the underside. Grows by the seaside in south-west England.

CLEMATIS, PURPLE
Clematis Viticella
The irritant poison protoanemonin renders this plant unfit for winemaking. Member of buttercup family ; flowers rose-purple.

28

CLOVER, ASLIKE
Trifolium Hybridum
All parts of the plant are possibly poisonous. Grows wild by roadsides throughout the country. The flowers are white, pinkish near the base, turning to brown late in August, after blossoming from June onwards. Grows to 60cm. (2ft.) tall.

CLOVER, RED
Trifolium Pratense
Could be a source of prussic acid, and all parts of the plant are best avoided for winemaking purposes.

CLOVER, ROUGH
Trifolium Scabrum
As red clover.

CLOVER, SOFT
Trifolium Striatum
As red clover.

CLOVER, STRAWBERRY
Trifolium Fragiferum
As red clover.

CLOVER, SULPHUR
Trifolium Ochroleucon
As red clover.

CLOVER, WHITE OR DUTCH
Trifolium Repens
As red clover, common in meadows throughout Britain. Grows to a height of 45cm. (18in.). The flowers are white or pinkish, appearing from May to October.

CLOVER, ZIG-ZAG
Trifolium Medium
As red clover.

COCKLEBUR, COMMON
Xanthium, Strumarium
Only the seeds are toxic, due to hydroquinone, which

produces paralysis, but the plant is best avoided. Its female heads consist of two flowers which develop into a rough, prickly seed case.

COCKLEBUR, SPINY
Xanthium Spinosum
As common cocklebur.

COLUMBINe OR AQUILEGIA
Aquilegia Vulgaris
This is a toxic plant, the seeds of which have been used to make a tisanes, but this is dangerous for children and can be dangerous for adults. To be found on chalky soil where shade is provided, and is, of course, cultivated. Grows to a height of 60cm. (2ft.), and the drooping violet-blue flowers have five petals, each with a separate backward-pointing lobe, appearing in May, June and July. The seeds are black and shiny.

CORN COCKLE
Agrostemma Githago
This plant is poisonous even after drying, particularly the seed. Fermentation does not affect the toxicity, and it is not for the use of winemakers ; the flower, in any case, is scentless. The seed, however, is bitter to the taste, and must not be countenanced for use in beer. The flowers are borne singly on long stalks, are red to purple in colour, with five petals, appear from June to August, and the resultant seeds resemble a curled-up caterpillar. This plant is no longer commonly found in cornfields. Grows to a height of one metre.

COTONEASTER
Cotoneaster Vulgaris
There are several varieties of this bushy, deciduous shrub, which grows up to 1 metre tall. Flowers pink. Poisonous.

COWBANE, OR WATER HEMLOCK
Cicuta Virosa
Probably the most poisonous of all plants in this country. The root can be mistaken for parsnip and the rootlets for potatoes, and contains in the yellow juice the greater amount of the convulsant poison cicutoxin, which is distributed

throughout the plant. Found in wet places in England and Wales. Grows to a height of one metre. The white flowers are in umbrella-like clusters up to 12 cm. across, appearing in July and August.

COW COCKLE
Saponaria
Contain poisonous saponins.

COW-WHEAT, COMMON
Melampyrum Pratense
Contains poisonous glycosides and is an extreme purgative. The flowers are yellow, or white with pink areas, and appear from June through to August. Found in woods and on moorland.

COW-WHEAT, CRESTED
Melampyrum Cristatum
Contains poisonous glycosides. The yellow, purple tinted flowers appear in clusters at the apex of the stem from June to September. Grows in woodlands in south-eastern England.

COW-WHEAT, FIELD
Melampyrum Arvense
Contains poisonous glycosides. The pink, with yellow patches, flowers grow at the top of the stem, and appear from June to September. Grows in cornfields in south-eastern England.

COW-WHEAT WOOD
Melampyrum Sylvaticum
Contains poisonous glycosides. The yellow flowers blossom at the base of the leaves from June to August. Grows in woodlands on mountain slopes in Scotland.

CROCUS, PURPLE
Crocus Purpureus
Doubtful for winemaking. The pale purple or white flower appears in March and April. Grows in meadows in south and eastern England.

CROTALARIA
Crotalaria
Poisonous.

CROWFOOT, CELERY-LEAVED
Ranunculus Sceleratus
The most poisonous of all members of the buttercup family, due to its luxurious growth, but not to the concentration of the irritant poison protoanemonin in each individual plant. The small yellow flowers appear from May through to September. The leaves are similar to those of celery. Grows up to 60cm. tall. Found in wet places.

CROWFOOT, CORT, OR CORV BUTTERCUP
Ranunculus Arvensis
The irritant poison protoanemonin renders this plant unfit for winemaking. Grows up to 60cm. tall. The yellow flowers appear in May, June and July. Found in the southern districts of England where the subsoil is chalk.

CROWFOOT, IVY-LEAVED
Ranunculus Hederaceus
The irritant poison protoanemonin renders this plant unfit for winemaking in the fresh state, but it could be used after drying. The petals are whitish. Found everywhere, in mud and shallow water.

CROWFOOT, RIGID-LEAVED, OR CIRCULAR-LEAVED
Ranunculus Circinatus
As Ivy Leaved Crowfoot. The whitish flowers appear from June to August. Found in slow-moving water.

CROWFOOT, RIVER
Ranunculus Fluitans
As ivy leaved crowfoot. The large whitish flowers appear from June to August. Found in fairly swiftly flowing water.

CROWFOOT, WATER
Ranunculus Aquatilis
As ivy leaved crowfoot. The flowers are whitish, and

appear in May and June. Grows in streams in south-east England where there is a chalky subsoil.

CROWN OF THORNS
Euphorbia Splendens
Poisonous. A member of the spurge family. A shrub grown as a pot plant for its scarlet-crimson flowers. Usually bare of leaves except at the flower head, or near.

CUCKOO-PINT, OR LORDS AND LADIES, OR WAKE ROBIN.
Arum Maculatum
All parts of the plant contain an irritant poison, calcium oxalate ; particularly the berries. The whitish root has an acrid, bitter taste. A common plant of hedges, woods and ditches, but local in Scotland. Grows to a height of 30cm. Bright, glossy green, sometimes purple-spotted, folded and pointed leaves appear in early spring. The yellowish-green-flower-leaf enfolds the purple-coloured rod-like flower, and when they wither away, a head of brilliant green berries are left, which turn bright red in July.

CUCKOO-PINT, GARDEN
Arum Italicum
All parts of the plant contain an irritant poison, calcium oxalate. The pale green sheathing leaf contains an orange spike flower, which appears in April and May. Found at the seaside in southern England.

CYCLAMEN, COMMON OR SOWBREAD
Cyclamen Hederifolium
Contains the poisonous glycoside cyclamin, and fermentation produces other poisons. Not for the winemaker. The purplish flowers appear in August and September. Grows in wooded areas of southern England.

CYAHAL
Minuartia Sedoides
Doubtful for winemaking. Grows in dense messy cushions, with minute whitish flowers which blossom from June to August. Found in Scotland on high ground.

DAFFODIL, PHEASANTS-EYE
Narcissus Poeticus
A purgative poison is found in the bulb, and the plant is best avoided for winemaking. The white flowers appear in May.

DAFFODIL, WILD
Narcissus Pseudonarcissus
Poisonous as Pheasants Eye Daffodil. The pale yellow flowers grow on stems up to 23cm. high, and blossom in March and April. Occurs in wet places.

DAHLIA
Species cultivated D. Coccinea, D. Coronata, D. excelsa
Tubers should be avoided for winemaking and little is known about the flowers ; should be regarded as doubtful.

DAPHE (upwards of 70 species)
The Daphne Mezereum is a shrub carrying white, pink, or purplish flowers in March before the leaves appear ; these flowers are profuse and sweetly fragrant ; they give rise to scarlet berries bunched along the stalks. Poisonous in all parts.

DARNEL
Lolium Temulentum
The plant, particularly the seeds, retains toxicity after drying and storage. A member of the rye grass family. Occurs on waste ground.

DEAD-NETTLE, HENBIT
Lamium Amplexicaule
This is a poisonous member of the dead-nettle family. The pink-and-crimson flowers appear from April through to August. Occurs on cultivated ground in southern and eastern England.

DELPHINIUM, GARDEN
Contains poisonous alkaloids.

DEVIL'S BIT
Succisa Pratensis
To be viewed with suspicion. The bluish-purple flowers grow on stems up to 45cm. high, and bloom from June to

September. Grows in damp meadows and in the fen country.

DOCK
Rumex
Should be treated with precipitated chalk to neutralise the poisonous soluble oxalates. There are at least a dozen species of this common and well-known weed.

DODDER, COMMON OR LESSER
Cuscuta Epithymum
A nasty poison. The red-coloured twining stems grow on gorse bushes and on heather, and the clusters of pink flowers blossom from July to September. Found in south and eastern England.

DODDER, FLAX
Cuscuta Epilinum
Plant is a parasite of flax ; twists up the flax in an anti-clockwise direction. Flowers yellow. Poisonous.

DODDER, GREATER
Cuscuta Europaea
The reddish twining stems grow on nettles and on hops, and the clusters of whitish flowers blossom from July to September. Grows in south and eastern England. Poisonous.

DOGBANE
Apocynum Venetum
Poisonous, due to the presence of ouabain, a virulent glycoside. A shrub growing to a height of half-a-metre, carrying pink flowers and having a delicate scent.

DOGS MERCURY
Mercurialis Perennis
This plant is poisonous due to the presence of mercuri-aline as the basic volatile oil content. The hairy stem has long-stalked flowers branching out from the leaf-base. The flowers grow up the higher part of the stalk, are yellowish, and blossom in March and April. Found in British woodlands.

DROPWORT, FINE LEAVED WATER, OR WATER HRMLOCK, OR DEAD MANS FINGERS
Oenanthe Crocata

All parts of the plant contain a convulsant poison which remains active after drying. The leaves must not be mistaken for celery ; nor the roots for parsnip. Fatalities have been recorded in such cases. Found throughout England in wet places. Grows to a height of 1.5 metres, and the white flowers appear in June, forming globular groups. The juice of the plant turns yellow on exposure to the air.

DROPWORT, FINE-LEAVED WATER
Oenanthe Aquatica

All parts of the plant contain a depressive poison, as distinct from convulsant as in *O. Crocata*, which remains active after drying and storage. The root must not be mistaken for parsnip, nor the leaves for celery. The whitish flowers appear from June through to September. Found in water.

DROPWORT, COMMON WATER
Oenanthe Fistulosa

Poisonous as fine-leaved dropwort. The stems are hollow, and the circlets of whitish flowers blossom from July to September. Grows in shallow water.

DUMBCANE
Dieffenbachia Picta and Seguine

Poisonous due to the presence of calcium oxalate. An indoor pot-plant having large leaves with various markings, including yellow or white spots; hybrids are common.

DUTCHMAN'S BREECHES
Dicentra Cucullaria

Poisonous. A garden plant. The variety Spectabilis is the well-known "Bleeding Heart."

ELDER, COMMON
Sambucus Nigra

The root and bark are to be avoided for winemaking, but the purgative action of the berries is not drastic, and they make a very acceptable wine.

ELDER, DWARF, OR DANEWORT
Sambucus Ebulus

The root and bark are to be avoided for winemaking, and the purgative action of the berries is not to be ignored. Grows to about 1 metere (3 ft.) in height. The pinkish-white flowers grow in groups of three, blossoming in July and August. Grows on waste ground.

ELDER, SCARLET-BERRIED
Sambucus Racemosa

The root and bark are to be avoided for winemaking, but the purgative action of the berries is not drastic. The creamy-white flowers appear in April and May. The scarlet, globose berries are more commonly met with in north-eastern parts of England.

ELEPHANT'S EAR
Haemanthus Albiflos

Poisonous due to the presence of calcium oxalate crystals, which are a burning irritant. An indoor plant with a dense head of white flowers above a double fan-like arrangement of thick, broad leaves.

FAT HEN
Chenopodium Album

Contains poisonous saponins, particularly in the roots. The small, greenish flowers grow on stalks from the base of the leaves, and appear in July and August. Found on cultivated land.

FERN, MALE
Dryopteris Filix-mas

To be avoided. Common through British Isles in woods, hedgerows and rocky places.

FIGWORT, COMMON
Scrophularia Nodosa

Contains a violent purgative poison. The stem is rectangular and the reddish flowers grow in groups, on stalks appearing from the base of the leaves, blossoming from June to September. Found in woods and hedges everywhere.

FIGWORT, GREEN-WINGED
Scrophularia Umbrosa
Contains a violent purgative poison. The reddish flowers grow in groups on thin stalks which grow from the base of the stem-leaves. Blossoms in July, August and September. Found in wet, shady places.

FIGWORT, WATER
Scrophularia Aquatica
Poisonous in all parts. Found around wet places everywhere. Grows to a height of 1.25 metres on a square stem. The brown and green flowers appear in July, August and September.

FIGWORT, YELLOW
Scrophularia Vernalis

Poisonous. The greenish-yellow flowers blossom in April, May and June. Grows in shady places.

FLAG, YELLOW OR CORN, OR IRIS, YELLOW
Iris Pseudacorus
Poisonous leaves and root, not destroyed by drying and storage. Grows up to 60 cm. tall. Two or three bright yellow flowers appear singly at the top of the stalk, sheathed in a leaf, in May, June and July. The leaves are long and pointed. Found in wet places everywhere.

FLAX, COMMON, OR LINSEED
Linum Usitatissimum
Should be boiled for $\frac{1}{4}$ hour, otherwise fermentation will produce prussic acid. The five-petalled flowers are blue, appearing in June and July. Grows up to 60 cm. high. The stems are smooth, the leaves long and narrow. Grown for its spherical seeds.

FLAX, PURGING
Linum Catharticum
All parts contain linamarin and the enzyme linamarase, producing prussic acid when fermented, and is to be avoided. Common on dry, chalky subsoil, pastures. Grows to a height of 15 cm., and the delicate white flowers appear from June to August.

FLAX, TOAD
Thesium

Contains poisonous glycosides. Includes 220 species. A perennial herb, growing up to 1 metre high. The narrow leaves resemble those of flax. The whole plant is yellowish-green in colour, the small flowers white. Bastard Toadflax (*T. Humifusum*) is a slender parasite living on roots on chalk subsoils.

FOXGLOVE OR DEAD MEN'S BELLS, OR THROATWORT
Digitalis Purpurea

Contains poisonous glycosides, including digitoxin, and remains poisonous after drying, storage and boiling; contained chiefly in the leaves, but all parts poisonous. There is a variety having white flowers, in addition to the common purple type. Grows everywhere in woods, and is commonly cultivated.

FOXGLOVE, STRAW
Digitalis Lutea

Poisonous—as *D. Purpurea*. A small yellow-flowered species, rust-coloured inside. The flowers are bell-shaped and form a spike.

FOXGLOVE, YELLOW
Digitalis Grandiflora

A larger variation of *D. Lutea* and just as poisonous.

FRITILLARY, OR SNAKE'S HEAD
Fritillaria Meleagris

This is a dangerous plant indeed, particularly the bulb, since it harbours the alkaloid imperialine, which has a fatal action on the heart. Grows up to 30 cm. high. Red, purple or yellow flowers growing singly or in pairs, in May, are spotted and with lines inside. Sparse, very long thin leaves. Bluey-green in colour. Cultivated plant which is found wild on wet land in the south of the country.

GLADIOLUS

Probably only the leaves and corms are dangerous, containing the glycoside iridin, which is still present after drying and storage. Best avoided.

GLADIOLUS, FIELD
Gladiolus Segetum
Contains a drastic irritant purgative, iridin, which remains after drying.

GLOBE FLOWER
Trollius Europæus
Contains the yellow volatile oil protoanemonin, which is at its maximum toxicity at flowering time, and which precipitates non-poisonous crystals when dried. Grows up to 60 cm. (2 ft.) tall. The almost globular, yellow, nine-petalled flowers blossom in June and July. Found in meadows.

GOLDILOCKS, WOOD
Ranunculus Auricomus
Contains the yellow volatile oil protoanemonin, which is at its maximum toxicity at flowering time, and which precipitates non-poisonous crystals when dried. Grows up to 30 cm. tall. The yellow, buttercup-like flowers appear from April through to July. Found in wooded areas.

GOOD KING HENRY
Chenopodium Bonus-henricus
Best avoided. Grows up to $\frac{1}{2}$ metre tall. The leaves are spear-head shaped, and the pale blue-grey flowers grow on diminishing short stalks to the head of the plant, blossoming in May through to August. Prefers a rich soil location.

GOOSEFOOT, FIG-LEAVED
Chenopodium Ficifolium
Best avoided. The stem is tinged with red. The greenish-white flowers grow in sparse clusters on stalks appearing from the leaf-base on the stem, and blossom in July, August and September. Grows on cultivated land, chiefly in southern and eastern England.

GOOSEFOOT, GREY
Chenopodium Opulifolium
Best avoided. Grows to 60 cm. (2 ft.) tall. The flowers appear in August and September. Found on waste ground.

GOOSEFOOT, MANY-SEEDED
Chenopodium Polyspermum

Best avoided. The stem is reddish and the flower stalks grow from the junction of the leaves with the stem. The minute flowers are in groups on the stalks, and appear from July to September. Appears on cultivated ground in southern England.

GOOSEFOOT, NETTLE-LEAVED, OR SOWBANE
Chenopodium Murale

Best avoided. The leaves are large compared with the short flower stalks branching from their base. The groups of flowers on each stalk are bluish, and appear in August. Grows on cultivated ground.

GOOSEFOOT, OAK-LEAVED
Chenopodium Glancum

Best avoided. The stem is reddish, as are the seeds. The small groups of flowers appear on stalks growing from the base of the leaves, blossoming in September. Found on rich land in southern England.

GOOSEFOOT, RED
Chenopodium Rubrum

Best avoided. The stem is reddish, and numerous flowers grow in closely knit groups on stalks growing from the base of the leaves, appearing in August. Grows in damp places in southern and eastern England.

GOOSEFOOT, STINKING
Chenopodium Vulvaria

Best avoided. The stem is bluish, and the flower spikes growing from the base of the leaves carry minute flowers in August and September. Appears on waste ground in southern England.

GOOSEFOOT, UPRIGHT
Chenopodium Urbicum

Best avoided. Flowers in August and September. The seeds are black. Grows on cultivated land.

GORSE, COMMON, OR FURZE
Ulex Europæus
To be viewed with suspicion. Grows to 2 metres tall. The yellow flowers appear in March, April and May. Appears on heaths.

GRAPE, OREGON
Mahonia Aquifolium
To be viewed with suspicion. The yellow flowers grow on a shrub, appear in April and May, and give rise to bluish-black globose fruit. Planted in woods.

GREASEWOOD
Sarcobatus
Poisonous due to the oxalate content.

GROUNDSEL, COMMON
Senecio Vulgaris
Can cause death by cirrhosis of the liver. The yellow flowers appear almost all the year round.

GROUNDSEL, STICKY
Senecio Viscosus
Will cause death by cirrhosis of the liver. The stem is stout, and the largish flowers appear in July, August and September. Grows on waste ground.

GROUNDSEL, WOOD
Senecio Sylvaticus
Poisonous—as Common Groundsel. The flowers appear in July to September in woods where there is a sandy soil.

HALOGETON
Halogeton
Contains very dangerous amounts of oxalates.

HAWTHORN, MAY OR WHITETHORN
You've probably made and imbibed many bottles of May-blossom wine without ill-effect. There is a poisonous element present in both the flowers and berries, but fermentation seems in some way to neutralise it, or the amounts

involved are negligible. Many medicines and herbal remedies contain small amounts of poisonous substances, of course.

HEDGE PINK, SOAPWORT, OR CROW-SOAP
Saponaria Officinalis
Contains poisonous saponins, which accounts for its descriptive alternative names, since it has been used for the same purposes as household soap. It grows up to 60 cm. high, with several stout stems to each root stock. The leaves taper at each end. The pink flowers carry a pleasant bouquet, and bloom in August in full heads at the top of the stems. Found everywhere, particularly in urban surroundings.

HELLEBORE, BLACK OR CHRISTMAS ROSE
Helleborus Niger
See Rose, Christmas.

HELLEBORE, FALSE
Veratrum
Poisonous alkaloids are present. The broadly oval leaves grow on a stout stem, and are distinctively pleated.

HELLEBORE, GREEN, OR BEAR'S-FOOT
Helleborus Viridis
All parts of this plant are poisonous, and these glycosides are not destroyed by storage or dehydration. Found in southern England on calcareous sub-soils in woods, and is cultivated. Grows to a height of 60 cm. (2 ft.) and the flowers are green, five-petalled, and appear in February, March and April.

HELLEBORE, STINKING, OR SETTERWORT
Helleborus Foetidus
All parts of the plant are poisonous, due to the presence of two glycosides which are unaffected by drying and/or storage. Found in southern England on chalk or limestone soils. The flowers are cup-shaped, pale green, drooping, and emit a foul smell; appearing from February to April.

HELLEBORINE, COMMON OR BROAD
Epipactis Helleborine
Best avoided in winemaking until more is known about

it. Common locally in woods in England and Wales. The flowers are purple in colour, grow in a loose spike up the stem, and appear from July to September.

HELLEBORINE, DARK-RED
Epipactis Atrorubens
Doubtful for winemaking purposes. The dark, reddish-purple flowers droop in echelon from the flower stalk, and appear in June and July. Found in limestone areas.

HELLEBORINE, LONG-LEAVED
Cephalanthera Longifolia
Doubtful. The white, tulip-like flowers appear from May to July. The leaves are grass-like. Found in woods on a chalky subsoil.

HELLEBORINE, MARSH
Epipactis Palustris
Doubtful. The pinkish-white flowers grow in a spike and spring individually from the stalk, appearing from June to August. Found in damp places.

HELLEBORINE, RED
Cephalanthera Rubra
Doubtful. The red flowers are white-tipped, tulip-like, and five or six grow up the stem on short stalks. Blossoms in June and July. Found in woods on a chalky subsoil.

HELLEBORINE, VIOLET
Epipactis Purpurata
Doubtful. The violet-coloured flowers appear in August and September. Grows in southern England's woods on a chalky subsoil.

HELLEBORINE, WHITE
Cephalanthera Damasonium
Doubtful. The white, tulip-like flowers, growing at intervals on the stem, appear in May and June. Found in southern England's woods on a chalky subsoil.

HEMLOCK
Conium Maculatum
Very poisonous, due to heavy concentration of alkaloids,

which are, however, destroyed by drying and heat. The
stems, covered with purple spots, are round and smooth,
growing 1½ metres or more tall. The flower clusters are
comparatively small, and appear in June and July. A smell
of mice is apparent when the plant is bruised. Similar to
parsley in appearance. Grows everywhere.

HEMLOCK, GROUND
Taxus

A poisonous member of the yew family, containing
toxic alkaloids. A needle-bearing evergreen shrub, recognisable from the yellowish undersides of the comparatively
broad needles.

HEMLOCK, WATER
Cicuta Virosa

Most of the poison is found in the root of this plant.
See Cowbane.

HEMP, MARIJUANA
Cannabis Sativa

The narcotic element of the whole plant is well-known.
Grows up to 2.5 metres in height. The sandpaper-like leaves are
divided into 5–7 narrow, serrated leaflets. The green flowers
appear on the extremities of the stems. It may surprise you
to learn that this plant grows wild in England, particularly
on rubbish dumps.

HEMP-NETTLE, COMMON
Galeopsis Tetrahit

To be avoided. The stem thickens where the leaf and
flower stalks grow. The purple and yellow flowers blossom
in July, August and September. Grows in woods and hedges.

HEMP-NETTLE, DOWNY
Galeopsis Segetum

To be avoided. The large, pale yellow flowers blossom
from July to August. Grows in sandy situations on cultivated ground.

HEMP-NETTLE, LARGE
Galeopsis Speciosa

To be avoided. The stems thicken at the point where
the leaves and flowers grow. The pale yellow, tipped with

violet, flowers appear in July, August and September. Grows on peaty soil.

HEMP-NETTLE, NARROW-LEAVED
Galeopsis Augustifolia
To be avoided. The pale pink flowers appear from July to September. Grows on cultivated ground.

HEMP-NETTLE, RED
Galeopsis Ladanum
To be avoided. The small, lilac-coloured flowers blossom from July to September. Grows both on cultivated and on waste ground.

HENBANE
Hyoscyamus Niger
All parts of the plant are very poisonous, and this toxicity is not reduced by boiling or by drying. The poison is hyoscyamine. Do not mistake this root for parsnip or chicory; it is the most poisonous part of all. Grows up to 60 cm. high and is a hairy plant. The smell is obnoxious. The pale yellowy-green or mustard-coloured flowers have purple veins, and appear on one side of the stem only, in July and August.

HENBIT
Lamium Urens
To be viewed with suspicion. The flowers are pink with a crimson petal, and appear from April to August. Grows on cultivated ground in southern and eastern England.

HERB PARIS
Paris Quadrifolia
Poisonous in all parts. Grows up to 30 cm. high, with four oval-shaped leaves up to 10 cm. long. The flowers are yellowish-green in colour, appearing in May, June and July, and the berry blue-black. Found in damp plantations on a chalk subsoil.

HOLLY
Ilex Aquifolium
The berries are poisonous. The small white flowers appear in April, May and June, to be followed by the berries.

HONEYSUCKLE
The berries are suspect; the flowers are safe.

HORSEBRUSH
Tetradymia
The liver injury caused by this plant could be severe.

HORSECHESTNUT
Aesculus
Poisonous, particularly new growth and the nuts.

HORSENETTLE
Solanum Auriculatum
A highly toxic member of the nightshade family. The flowers are violet in colour and appear in groups of four or so with a leafy base, at the head of the plant. There are other species of this family, which are grown as pot plants.

HORSE-RADISH
Armoracia Rusticana
The volatile oil is a strong irritant when used to excess; well-known as a sauce, of course, and therefore acceptable if the taste is not overpowering. Cultivated.

HORSETAIL
Equisetum
Poisonous in all parts. Favours damp, acid soils everywhere. The name of the plant is descriptive—a main stem carries numerous long green stalks.

HOUND'S-TONGUE
Cynoglossum Officinale
An irritant poison is present in all parts. Found on grassed places. Grows up to 1 metre tall on a stout, hairy, branching stem. The dull-red flowers appear in May, June and July.

HYACINTH, NEGLECTED
Muscari Neglectum
All parts poisonous, particularly the bulbs. Similar to bluebell, purple-flowered.

HYACINTH, TASSEL
Muscari Comosum
All parts of the plant are poisonous, particularly the bulbs. The buds are violet, the lower hanging mature flowers brown, and the upper upright flowers reddish. The flowers are blue in Grape Hyacinth (*M. Racemosum*).

HYDRANGEA
Hydrangea
Contains poisonous cyanides, but the characteristics of such poisoning do not materialise, which makes for additional danger.

IRIS, PALE-BLUE
Iris Spuria
Contains the glycoside iridin, which remains after drying and storage—best avoided. Flowers whitish, with violet overtones, blossoming in June and July. Grows chiefly in Lincolnshire.

IRIS, STINKING, OR GLADDON, OR ROAST BEEF
Iris Fœtidissima
Contains the poisonous glycoside iridin in all parts, which is not removed by drying or storage. The flowers appearing in June are blue or whitish in colour, on dark green stems carrying dark ever-green leaves. The latter emit an obnoxious smell when bruised. Found only in southern England in shaded woodland on dry, chalky subsoils.

IRIS, YELLOW, OR FLAG, YELLOW
Iris Pseudacorus
Contains a purgative and irritant glycoside, which remains after drying and storage. Found in wet places everywhere. Grows up to 60 cm. high, with two or three large yellow flowers at the top. The pale greenish-brown seeds are contained in a green sheath.

IVY
Hedera Helix
The leaves and berries are poisonous.

Ivy, Ground
Glechoma Hederacea
Poisonous, but small amounts have been used in the past to flavour and clarify beer. Two handfuls in a pint of boiling water has been used as a tonic tisane at the rate of three wineglassfuls per day. Common in all parts of Britain. The square stems are trailing, and root as they go. The leaves are kidney-shaped and the flowers are a brilliant purplish-blue, appearing throughout the summer.

Jack-in-the-Pulpit
Arisaema Ringens
The crystals of calcium oxalate contained in this plant, particularly in the roots, are dangerous. The same applies to Arisaema Triphyllum, or Indian Turnip. The flowers are recognisable by their translucent stripes.

Jasmine, Yellow
Gelsemium Sempervirens
The flowers contain poisonous alkaloids, and the roots the toxic bitter alkaloid gelsemine. An evergreen shrub with bright yellow, bell-shaped flowers.

Jonquil
Narcissus
The bulbs contain poison and the whole plant is best avoided.

Juniper
Juniperus Communis
Use only the berries—the leaves are suspect. This shrub is found on chalk and limestone subsoils.

Kale
Best used in amounts not exceeding a normal meal intake. (See Cabbage).

Kalmias
Poisonous, and the flowers are to be avoided.

Knotgrass, Knotweed, or Wireweed
Polygonum Aviculare
Contains an irritant poison. Grows prostrate and the

pinkish flowers appear in July and August. Grows everywhere.

KOHLRABI

Best used in amounts not exceeding a normal meal
intake. (See Cabbage).

LABRADOR TEA
Ledum Groenlandicum

Contains the poisonous andromedotoxin. Evergreen
growing to height of less than 1 metre. Stem and lower side
of leaves are felt-like and rust-coloured.

LABURNUM, COMMON, OR GOLDEN RAIN
Laburnum Anagyroides

All parts of this tree are extremely poisonous, particularly the bark and seeds. A cultivated tree, seldom seen
growing wild. The well-known golden-yellow flowers appear
in May and June.

LANTANA
Lantana Camara

Poisonous, with particular mal-effects on the liver.
A pot plant, of which there are about 160 species. It is a
small shrub, and the flowers are orange coloured at first,
turning to yellow and then red.

LARKSPUR
Delphinium Ajacis

The seeds and foliage are poisonous, but the poisons
are removed by drying and storage. Cultivated. Found
wild in cornfields of eastern counties. Grows to 60 cm. tall.
The dark blue flowers appear in June and July.

LARKSPUR, FORKING
Delphinium Consolida

Contains poisonous alkaloids. The five-petalled mauve
flowers appear in June and July on stalks up to half a metre
tall.

LAUREL, CHERRY
Prunus Laurocerasus

All parts of the shrub are poisonous. Evergreen shrub
found in ornamental gardens. The white flowers are sweet-

scented, which could attract the experiment-minded, as could the black cherry-like fruits. Grows to a height of 6 metres.

LAUREL, MOUNTAIN
Kalmea Latifolia
Contains the poisonous andromedotoxin. Shrub usually about 3 metres tall. Attractive pink flowers with unusual protuberances on the petals.

LAUREL, SPURGE OR WOOD
Daphne Laureola
Very poisonous in all parts. A shrub growing up to 1½ metres tall, with few branches, having evergreen leaves at their tips. The flowers are greenish-yellow and sweet-scented, appearing in February through to May. The fruit is an oval-shaped black berry. Found in English woodlands on chalk subsoils.

LECHUGUILLA
Agave
The poison content acts particularly on the liver. There are about 300 species which are cultivated as indoor ornamental plants. The flowers grow in groups near the top of a long stalk, the leaves forming the base of the plant.

LETTUCE, WILD, OR ACRID
Lactuca Virosa
Doubtful. The small yellow flowers, on delicate stalks up to 1 metre tall, appear in July and August. Grows near the sea in eastern England.

LILAC
All parts poisonous.

LILY, BLOOD
Haemanthus
Contains poisonous alkaloids. Cultivated in slightly heated greenhouses. Grows to a height of 30 cm. (1 ft.) with spherical scarlet flowers in clusters, which are followed by strap-shaped foliage.

LILY, CLIMBING OR GLORY OR FLAME
Gloriosa Rothschildiana
Contains poisonous alkaloids. Cultivated in slightly-heated greenhouses. Climbs by means of tendrils at the leaf tips.

LILY, CRINUM
Crinum Kirkii
Poisonous, particularly the bulb. This flower is red in colour, but there are several varieties ranging from white through to red. Grown as a tub plant.

LILY-OF-THE-VALLEY
Convallaria Majalis
Contains two poisonous glycosides, and all parts are poisonous. Grows in woodlands where there is a chalky subsoil, and is cultivated. The bell-like, white, scented flowers appear in May.

LOBELIA, ACRID
Lobelia Urens
Doubtful for winemaking; contains an alkaloid similar to nicotine, which is acrid to the taste. Found on moorland and in woods in Devon, Cornwall, Hampshire and Sussex. The blue flowers grow alternately on the flower stem, which reaches a height of 60 cm.

LOBELIA, WATER
Lobelia Dortmanna
Contains poisonous alkaloids. The pale lilac-coloured small flowers grow sparingly on the stem, and appear in July. The leaves are submerged. Grows in lakes having a gravel bottom.

LOCO
Astragalus
Poisonous. Herbs or shrubs the flowers of which are followed by pods.

LOCUST, BLACK
Robinia
Poisonous, particularly the seeds in the long, flat pods, but also the bark and other parts are dangerous. A tree.

LOUSEWORT
Pedicularis Sylvatica
Contains poisonous glycosides. Grows up to 15 cm.
tall. The pale mauve flowers appear from April to July.
A common plant on heaths and on hilly pastureland.

LUCERNE
Medicago Sativa
To be avoided. Grows up to 60 cm. (2 ft.) tall. The
purple flowers grow in clusters and appear from April through
to August. Cultivated, and also found as an escape.

LUPIN
Lupinus Angustifolius (Blue)
Lupinus Luteus (Yellow)
Contains poisonous alkaloids in all parts. Cultivated.

MAHOGANY, MOUNTAIN
Cercocarpus
Contains cyanogenetic glycosides, a potent poison.

MANCHINEEL
Hippomane
Contains caustic poison. A small tropical tree.

MALLOW, COMMON
Malva Sylvestris
Suspect. The five-petalled reddish-purple flowers appear
from June to September. Grows everywhere.

MALLOW, DWARF
Malva Neglecta
Suspect. Grows prostrate. The small, very pale purplish
flowers blossom from June to September. Grows in southern
England.

MALLOW, LONG-HAIRY OR HISPID MARSH
Althæa Hirsuta
Suspect. The hairy stem grows up to half a metre tall.
The five-petalled pale mauve flowers appear in July and
August. Grows in south-east England.

MALLOW, MARSH
Althæa Officinalis
Suspect. Grows up to 1 metre tall. The five-petalled, pale mauve flowers grow in clusters, blossoming in August and September. Grows on seaside marshes.

MALLOW, MUSK
Malva Moschata
Suspect. Grows up to 60 cm. (2 ft.) tall. The pale mauve or pinkish flowers bloom in July and August. Found on dry embankments.

MALLOW, SMALL
Malva Pusilla
Suspect. The small pale mauve flowers appear from June through to September.

MANGOLD OR MANGEL WURZEL
Beta Vulgaris
Do not use until after Christmas, to be on the safe side, and use precipitated chalk to neutralise the oxalates.

MARIGOLD, MARSH, OR KING CUP OR MOLLYBLOBS
Caltha Palustris
Poisonous. Grows in wet meadows and marshes everywhere. The golden-yellow flowers appear in March, April and May.

MAYAPPLE
Podophyllum
Poisonous. These plants consist of one or two leaves, the last-mentioned plant carrying a short-stalked white flower, which gives rise to an ovoid, blotched-yellow fruit.

MEADOW-RUE, COMMON
Thalictrum Flavum
Suspect, even after drying and storage. Grows up to 1½ metres tall on a stout stem. The small yellowish flowers grow in groups, several groups to each stalk, and flowers in June and July. Found in damp places.

MEDICK, BLACK
Medicago Lupulina
Doubtful. The groups of three leaves are similar to those of clover. The plant grows prostrate. The small yellow flowers appear in tightly-knit groups from May through to August. The pods are black in colour. Found everywhere on pastureland.

MEDICK, BUR OR SMALL
Medicago Minima
Doubtful. A prostrate plant with very small clover-like leaves. The minute yellow flowers blossom from May to July. Found in south-eastern England in sandy places.

MEDICK, HAIRY OR FIMBRIATE OR TOOTHED
Medicago Polymorpha
Poisonous, and remains so when dried and stored. The prostrate stem carries clover-like leaves and yellow flowers, which blossom from May through to August. Grows on the south and east coasts of England.

MEDICK, SICKLE OR YELLOW
Medicago Falcata
Doubtful. The plant usually grows prostrate, and the leaves are narrow, in threes. The yellow flowers blossom in June and July. Grows in eastern England below the Wash on dry sand and gravelly soils.

MELILOT, OR SWEET YELLOW CLOVER
Melilotus Officinalis
All parts poisonous. Cultivated, or found as an escape. Grows to 1 metre tall, with erect branched stems. The yellow flowers grow drooping in spikes in June and July.

MERCURY, ANNUAL, OR WILD SPINACH
Mercurialis Annua
The volatile oil in all parts of the plant is poisonous. Common on cultivated land in south-east England. Grows to a height of 30 cm. The green flowers bloom from July to September. Branched stalks distinguish it from Dog's Mercury.

Mercury, Dog's or Herb
Mercurialis Perennis

The volatile oil in all parts of the plant is poisonous, but dissipates in time, and is removed altogether by drying or boiling. Common in woods and shaded places as far north as the south of Scotland. Grows to a height of 30 cm. (1 ft.) with branched stems, and the very small green flowers appear in April and May, before the leaves.

Mezereon, or Bay Tree, Dwarf, or Olive, Spurge
Daphne Mezereum

Contains a poisonous acid in all parts, which is not destroyed by drying and storage. Grows as a shrub up to 1 metre (3 ft.) high. The flowers are pink or purple, and appear between February and April before the leaves, which grow up to 8 cm. long on the ends of the branches. The fruit should not be mistaken for red currants. A cultivated shrub which also grows wild in woodlands in southern England.

Milkweed, Labriform
Asclepeas Syriaca

An extremely toxic plant. Cultivated as a pot plant. The leaves are large and oval in shape. The flowers are spherical and violet in colour. The flat seeds occur in conjunction with a mass of brittle fibres. There are a great number of species in this family.

Millet

Should be boiled for at least $\frac{1}{4}$ hour, or fermentation will produce prussic acid. A variety has been used in beer brewing in the past. I have personally made and imbibed this wine.

Mistletoe
Viscum Album

The berries are slightly poisonous. The cultivated and wild plants carry their greenish-yellow flowers in February, March and April.

Mistletoe, American
Phoradendron Flavescens

Contains toxic amines. Similar in appearance to our own mistletoe.

MOONSEED
Menispermum
Poisonous. A vine of which the "grapes" can be recognised by the single, large, crescent-shaped seed, as opposed to the many seeds to be found in edible grapes, and the absence of seed in edible seedless grapes.

MUGWORT
Artemisia Vulgaris
Doubtful. Has been used in the past to flavour beer, and some authorities describe a tisane made from it. The strong, reddish, grooved stem grows up to 1 metre tall. The leaves are whitish underneath. The greenish-pink flowers grow up branched stalks, and blossom from July to September. Found on waste ground.

MUGWORT, BRECKLAND
Artemisia Campestris
Doubtful. Grows up to half a metre tall with a thin branched stem and linear leaves. The greenish-pink flowers grow up the stalks, and blossom in August and September. Found on heaths with a sandy subsoil in eastern England.

MULLEIN, COMMON
Verbascum Thapsus
Contains poisonous glycosides. The five-petalled yellow flowers grow in a tight cone-shape at the apex of the thick stalk, which is clothed with leaves around it lower down. Flowers from June to August. Grows on dry soil everywhere.

MULLEIN, DARK
Verbascum Nigrum
Contains poisonous glycosides. The dark green leaves distinguish it from Common Mullein. Flowers from June through to September. Grows on dry, chalky-subsoil banks.

MULLEIN, HOARY
Verbascum Pulverulentum
Contains poisonous glycosides. The leaves are clothed with white hairs as distinct from Common Mullein. Blossoms in July and August. Found locally in eastern England.

MULLEIN, LARGE-FLOWERED
Verbascum Thapsiforme
Contains poisonous glycosides. Similar to Common Mullein, but the flowers are larger. Found on dry banks.

MULLEIN, MOTH
Verbascum Blattaria
Contains poisonous glycosides. The flowers are solitary as distinct from those of Common Mullein, and appear from June through to September. Grows on waste ground in southern and eastern England.

MULLEIN, ORANGE
Verbascum Phlomoides
Contains poisonous glycosides. The yellow flowers appear from June to August. Grows on waste ground in southern England.

MULLEIN, PURPLE
Verbascum Phœniceum
Contains poisonous glycosides. The spikes of flowers can grow up to 2 metres tall.

MULLEIN, WHITE
Verbascum Lychnitis
Contains poisonous glycosides. The flowers are whitish and the leaves have white hairs on the underside, as distinct from Common Mullein. Blossoms in July and August. Found in southern England on chalk subsoils.

MUSTARD, BLACK
Brassica Nigra
The seed pods are suspect. The small yellow flowers are grouped at the extremity of the stalk and appear from June to August. Found at the seaside in south-western districts.

MUSTARD, WHITE
Sinapis Alba
The seeds and pods are poisonous. Common in cultivated land. Grows to a height of 1½ metres. The four-petalled flowers appear in June, July and August.

NARCISSUS
A purgative poison is present in all parts of the plant.

NERINE
Nerine Undulata
Poisonous. A pot plant grown in cold greenhouses. Grows to a height of 60 cm. with pale pink flowers which blossom from September to November. There are other varieties, but most have reddish flowers; which grow from bulbs.

NIGHTSHADE, AMERICAN, OR POKEWEED, OR PIGEONBERRY
Phytolacca Americana
All parts of the plant are poisonous. It grows up to 2 metres (6 ft.) in height. The stalk is thick and the flowers form a spike at its extremity. The berries are black when ripe, and emit a red-coloured juice. This is a garden plant, sometimes found as an escape.

NIGHTSHADE, BLACK OR GARDEN
Solanum Nigrum
Poisonous in all parts. Grows up to 30 cm. in height. The stalk is branched. The small five-petalled, star-like flowers are white in colour and grow in clusters from July to October. The ripe berries are sometimes black, sometimes red, and sometimes green, according to the species. Found everywhere as a weed, but more particularly in southern England, and at the seaside. There is, however, a species (*Solanum Nigrum* var. Guineense) which is also known as Huckleberry, and is safe for winemaking.

NIGHTSHADE, DEADLY, OR DWALE, OR BELLADONNA, OR BANEWORT
Atropa Belladonna
All parts of this plant are poisonous, particularly the leaves and roots, due to the presence of alkaloids which are not removed by dehydration or boiling. Atropine and hyoscine are the drug derivatives. Found in woods on chalky soil in England and Wales, particularly in the south, and locally elsewhere on waste ground. Grows to a height of 1½ metres, with unequal-sized dark green leaves in pairs,

and the dull purple bell-shaped flowers appear from June to August, to be followed by the green, then red, then glossy black berries, which are about 12 mm. in diameter.

OAK, COMMON OR BRITISH
Quercus Robur
The whole tree is rich in poisonous tannic acid, and the winemaker will only need, and use, small amounts of the leaves to give his wine the typical tannin "bite". This is the oak tree found on clay soils in England and southern Scotland.

OAK, CORK
Quercus Suber
Use only a modicum of the leaves for their tannin.

OAK, DURMAST
Quercus Petraea
Use only a small quantity of the leaves for their tannin. Native to this country.

OAK, EVERGREEN OR HOLM
Quercus Ilex
Doubtful for winemaking. Mostly found in parks.

OATS
Always boil for ¼ hour to be on the safe side, but the danger, if any, of nitrate poisoning is more probable in the case of American-grown cereals, which can be rich in poisonous nitrates.

OLEANDER
Nerium Oleander
The leaves and wood are poisonous, and the nuts are lethal. This is a shrub with leathery leaves, often grown in tubs. The flowers are normally pink, but there are white, yellow, red and multi-coloured varieties.

ONION
Toxic if taken in large quantities over a short period, since it is anemia-inducing. Not likely to cause trouble for the winemaker if imbibed as wine in amounts not exceeding a normal meal intake.

ORACHE, COMMON
Atriplex Patula
Poisonous if taken in large quantities. Cultivated. Flowers in August and September.

ORACHE, FROSTED
Atriplex Laciniata
Poisonous in large quantities. The whole plant is silvery in colour. Flowers in July, August and September. Found at the seaside.

ORACHE, HALBERD-LEAVED
Atriplex Hastata
Poisonous in large quantities. The leaves are almost triangular in shape. The stem is sturdy and the greyish flowers tipped with pink. Blossoms in August and September. Found at the seaside.

ORACHE, SHORE
Atriplex Littoralis
Poisonous in large quantities. The flower stalks are more numerous than in Common Orache, and the leaves are almost linear. Blossoms in July and August. Found on the seashore.

PAEONY
Paeonia Mascula
Doubtful. Grows up to 60 cm. (2 ft.) tall. The flower and the stalk are red. Blossoms in May and June. Found in south-west England.

PARSLEY, CAMBRIDGE
Selinum Carvifolia
Suspect. The stem is ridged. Flowers from July to October. Found in damp places in Cambridgeshire.

PARSLEY, COW
Anthriscus Sylvestris
Suspect. The leaves are fern-like. The flowers, carried in groups at the top of a delicate stalk, appear in April, May and June. Grows almost everywhere.

PARSLEY, FOOL'S OR LESSER HEMLOCK OR DOG'S PARSLEY
Aethusa Cynapium

The leaves have been mistaken for parsley, and the roots for small turnips or radishes; which in conjunction provide a means of recognition, as does the repulsive odour of the bruised plant. The two poisonous alkaloids are rendered harmless by drying and storage. This is a common weed growing not more than 60 cm. tall, and carries small white flowers; parsley flowers are yellow, and radish flowers are whitish-mauve, four-petalled. The flowers appear in July and August.

PARSLEY, MILK
Peucedanum Palustre

Doubtful. The sturdy stem is hollow, and the pale-bluish flowers grow in groups of sparse clusters, appearing from July to September. Found in wet situations, particularly on marshland.

PARSNIP, WATER
Sium Latifolium

Doubtful. The hollow stem grows up to 2 metres tall with luxuriant, longish, finely serrated leaves. The whitish flowers grow in separate-stalked clusters from the top of the stem, and appear in July and August. Found on marshland in eastern England, particularly in the fen district.

PARSNIP, WILD
Pastenaca Sativa

Doubtful. The hollow stem carries long flower stalks, with a few offshoots at the top carrying groups of small yellow flowers, which blossom in July and August. Found on chalk subsoils in the southern half of England.

PASQUE FLOWER
Pulsatilla Vulgaris

Poisonous leaves and stalks. The flower is a dull purple in colour and appears in April and May. Grows to a height of 30 cm. Found rarely, and only in England within the British Isles. Cultivated.

PEA, BLACK OR VETCH, BLACK BETTER
Lathyrus Niger
The seeds are poisonous, and probably the whole plant. Seldom found in Britain.

PEA, MARSH
Lathyrus Palustris
The seeds are poisonous, and probably the whole plant. Grows up to 1 metre tall. The leaves are pointed-oval. The purplish flowers bloom in June, July and August. Found on very wet ground.

PEA, EVERLASTING, OR WILD
Lathyrus Sylvestris
The seeds are dangerous and the whole plant is poisonous. Grows up to 2 metres tall, with long, pointed-oval leaves. The pinkish-yellow flowers blossom from June to August. Found in woods and in wet places at the seaside.

PEA, ROSARY, OR BEAN, PRECATORY
Abrus
Poisonous. The greater part of the 9 mm. long seed is scarlet-coloured, and the rest black; it is contained in a pod. A house plant.

PEA, SEA
Lathyrus Maritimus
The seeds are poisonous, and probably the whole plant. The leaves are oval-shaped, and the reddish-purple flowers blossom from June to August. Found on pebbly seashores.

PEA, SWEET
Lathyrus Odoratus
The seeds are poisonous, and probably the whole plant. Cultivated.

PEACH
The kernel yields prussic acid, and is best removed prior to winemaking.

PEACHWORT, PERSICARIA, OR REDSHANKS
Polygonum Persicaria
Contains an acrid poison.

The pointed-oval leaves are dark-stained, and the flower stalk is branched, carrying pink flowers grouped like a head of grain, and blossoming from June to September.

Found on damp, cultivated ground.

PEAR
The pips yield prussic acid, and you may prefer to core your pears before use for winemaking, particularly when making bulk quantities.

PENCIL TREE
Euphorbia
Poisonous.

An ornamental plant.

PEONY
Poisonous.

Cultivated.

PEPPER, WATER OR SMARTWEED OR PERSICAIA BITING
Polygonum Hydropeper
Contains a sharp, acrid juice, and is not advisable for wine-making purposes.

Grows up to 60 cm. tall, but bends over at the top. The flowers are greenish, growing at intervals up the stalk, blossoming from July to September.

Found in damp situations.

PERSICARIA, PALE
Polygonium Lapathifolium
Poisonous.

The greenish-white flowers grow in dense groups on stalks branching from the stem and in the shape of an ear of corn. They blossom from June to September.

Found on damp ground.

PHEASANT'S-EYE, SPRING
Adonis Vernalis
Poisonous.

Grows up to 30 cm. high, with 5-8 scarlet petals, having a dark spot nearer the stalk.

Found only in warmer districts.

PHEASANT'S-EYE, SUMMER
Adonis Aestivalis

Poisonous. as A Vernalis

PHILODENDRON
Philodendron

Contains grossly irritant oxalate crystals.

An ornamental pot plant, of which there are 250 species. Often having broad, heart-shaped leaves.

PIERIS
Pieris

Contains poisonous andromedotoxin.

PIMELEAS
(About 80 species)

Poisonous. Shrub-type plants with serried rows of very small leaves terminated by groups of flowers.

A cultivated plant.

PIMPERNEL, BOG
Anagallis Tenella

Poisonous.

The trailing stem carries disc-like leaves and the delicate upright flower stalks carry pale pink bell-shaped blossoms from June to August.

Found on wet peat in the counties south of the Wash.

PIMPERNEL, COMMON OR SCARLET
Anagallis Arvensis

Poisonous.

The five-petalled scarlet flowers grow on delicate stalks from the frailing stem. The leaves are stalkless, pointed-oval in shape.

Found on cultivated ground.

PINKS

Contains poisonous saponins, but in amounts not likely to be dangerous in the quantities used for winemaking.

Cultivated.

PINK, CHEDDAR
Dianthus Gratianopolitanus
Contains poisonous saponins.

The pink flowers blossom in June and July.

Grows wild on chalky subsoils in Somerset.

PINK, DEPTFORD
Dianthus Armeria
Contains poisonous saponins.

The crimson flowers blossom in July and August.

Grows wild on sandy soils.

PINK, FEATHER
Dianthus Plumarius
Contains poisonous saponins.

The pink flowers appear in June, July and August.

The ancestor of the cultivated pink.

PINK, GERMAN
Dianthus Carthusianorium
Contains poisonous saponins.

PINK, MAIDEN
Dianthus Deltoides
Contains poisonous saponins.

The darker pink petals have a touch of red at their base. Blossoms from June to September.

Found in dry situations.

PLUM
The kernel yields prussic acid, and is best removed prior to winemaking.

POINCIANA OR BIRD OF PARADISE
Poinciana
The pods are poisonous.

A pot-grown shrub having large yellow flowers formed like the sweet pea, and giving rise to pods up to 10 cm. long.

POINSETTIA
Euphorbia Pulcherrima
Contains a caustic poison.

A house plant, in the form of a shrub, with scarlet flowers.

POKEWEED
Phytolacea

Poisonous, particularly the root.

The inconspicuous, small, white or greenish flowers, on thick purplish stems, give rise to purplish berries.

POPPY, BRISTLY
Papaver Hybridum

Poisonous in all parts.

The four red petals are well separated. Blossoms in June and July.

Found on sandy soil in south-eastern England.

POPPY, HORNED RED
Glaucium Corniculatum

Poisonous in all parts.

POPPY, HORNED, OR SEA
Glaucium flavum

Poisonous in all parts. Found on shingle foreshores. The yellow flowers appear from June through to October.

POPPY, ICELAND
Papaver nudicaule

Poisonous in all parts. Cultivated. Yellowish flower appearing from June through to August.

POPPY, LONG-HEADED
Papaver dubium

Poisonous in all parts. The brownish-red flowers appear in June and July. Grows in cornfields.

POPPY, OPIUM OR WHITE
Papaver somniferum

The seed vessels contain opium, which is a source of morphine; the tincture of which is commonly known as laudanum, and a derivative of which is heroin. All parts are poisonous. Can be seen occasionally in the eastern counties of England. The large flowers may be white, pink, or lilac in colour, and appear from May to June.

POPPY, PALE
Papaver argemone
Poisonous in all parts. The four red petals are well separated. Blossoms in June and July. Grows in dry cornfields.

POPPY, RED
Papaver rhœas
A large quantity is poisonous. The dried petals are used in the preparation of red-coloured and bitter medicines. This poppy does not contain opium. Can be seen in cornfields. The scarlet flowers appear from June to September.

POTATO
Solanum tuberosum
Poisonous solanines are present in the stalks, leaves, green shoots, and in any tubers which have become green due to exposure to light.

POTATO, SWEET
Ipomoea batatas
Only in exceptionally nitrate-rich soil does this plant become dangerous to man, and then only if consumed in relatively large amounts.

PRIMROSE
Primula vulgaris
A surprising one to find in this list. The plant *does* contain poison, so we list it as "doubtful," but the petals alone have long been used for winemaking without known ill effect. The pale yellow flowers appear from February to May. Found on wooded slopes.

PRIVET, COMMON
Ligustrum vulgare
All parts poisonous, particularly the berries.

RADISH, WILD
Raphanus raphanistrum
The seeds are poisonous. Common on cultivated ground. Grows to a height of 60 cm. The yellow or white flowers are four-petalled, and appear from May through to August.

RAGWORT, BROAD-LEAVED
Senecio fluviatilis
Poisonous. Grows to one metre in height. The yellow flowers appear in August. Found near running water.

RAGWORT, COMMON, OR BEN WEED, OR CURLEY DODDIES, OR ST. JAMES'S WORT, OR STAGGERWORT
Senecio jacobœa
Poisonous alkaloids in all parts remain active after drying and storage. Grows up to one metre (3 ft.) in height. The erect grooved stems are red-coloured below, and green and branched above. The light green leaves are ragged. The large brilliant-yellow flower clusters are flat-topped, and appear from June to October. Can be seen almost everywhere in Britain, mainly on pastureland.

RAGWORT, GREAT FEN
Senecio paludosus
Poisonous. Grows to almost two metres in height. The leaves are long, serrated, and pointed. The yellow flowers appear from May to July. Found in the fen district.

RAGWORT, HOARY
Senecio erucifolius
Poisonous. Grows to one metre in height. The yellow flowers appear in July and August. Found on clay soils in south-eastern England.

RAGWORT, MARSH
Senecio aquaticus
All parts poisonous, the responsible alkaloids being unaffected by storage or dehydration. Grows to a height of one metre in wet meadows. The flowers are bright yellow with a yellow centrepiece ringed with orange, and appear from July to August.

RAGWORT, OXFORD
Senecio squalidus
Poisonous in all parts. Grows to 30 cm. tall on a stout stem. The yellow flowers appear from June through to September. Found on embankments.

RAGWORT, SPREADING
Senecio erraticus
Poisonous.

RAMSONS, OR GARLIC, WOOD
Allium ursinum
Doubtful, best avoided. The six-petalled, white, star-like flowers appear in April, May and June. The leaves are pointed, elliptical. The smell is unmistakable. Found in woods on damp ground.

RAPE
Brassica napus
Best avoided, probably poisonous in all parts. Cultivated by farmers. In May and June the brilliant yellow flowers appear on stems up to 60 cm. (2 ft.) tall.

RAUWOLFIA
Rauwolfia serpentina
Poisonous, due to the alkaloids in the roots. A native plant of India; grows as a small shrub. There are varieties in most tropical countries, and it can be found in hot-houses.

REDLEG, OR PERSICARIA, COMMON
Polygonum persicaria
To be viewed with suspicion. The pink flowers blossom from June through to September. Grows on cultivated ground having a clay subsoil.

RED-RATTLE
Pedicularis palustris
Poisonous. This pink-stemmed, pink-stalked, pink and crimson flowered plant blossoms from May through to September. It grows up to half a metre tall. Found in wet places on moorland.

RHODODENDRON, COMMON, OR PONTIC
Rhododendron ponticum
Poisonous in all parts, particularly the flowers and leaves, due to the presence of glycosides.

RHUBARB
Rheum rhaponticum
The leaves are poisonous and must not be used. The

stalks contain an excess of oxalic acid, which it is often recommended should be neutralised with precipitated chalk, but in practise we have never heard of anyone coming to harm by drinking rhubarb wine even when it has been made without this being done.

ROSE, CHRISTMAS, OR BLACK HELLEBORE
Helleborus niger
All parts of the plant are poisonous, due to the presence of glycosides helleborein and helleborin, which are not affected by drying and/or storage, but prolonged boiling has some effect; even so, this plant should be avoided by the wine-maker. Popular as a garden plant throughout Britain. Grows to a height of 60 cm. (2 ft.). Has very dark green leaves, and the five-petalled white or whitish-pink rose appears from December to February.

ROSE, GUELDER
Viburnum opulus
To be viewed with suspicion. The flowers are white, five-petalled, and appear in June and July. Grows in damp woods and hedges.

ROWAN, OR MOUNTAIN ASH
The berries yield prussic acid, but may be acceptable when used in the small quantities required for winemaking. A sweet wine will be much safer than its dry counterpart; this fact is said to have saved Rasputin's life when prussic acid was added to his wine on one occasion.

RUSH, BLUE OR HARD
Juncus inflexus
Poisonous in all parts. Grows up to 60 cm. (2 ft.) high in dense tufts. Commonly found on damp pastureland throughout Britain.

RUTABAGA
Senecio brassica
Dangerous if taken in large amounts in conjunction with an iodine-deficient diet. One side is covered with cream-coloured down.

RUSTY-LEAF
Menziesia caprulea

Contains the poisonous andromedotoxin. Bushy shrub rising only to 15 cm. in height.

RYE

Should be boiled for ¼ hour at the least, particularly when from American sources.

SACHUISTA OR BEAR-GRASS
Nolina

Can cause liver-injury. An ornamental with a dense mass of stiff, pointed leaves.

SAFFRON
Crocus sativus

Doubtful. The source of a yellow dye, for which it is widely cultivated. A crocus-type flower.

SAFFRON, MEADOW, OR AUTUMN CROCUS, OR COLCHICUM, OR NAKED LADY
Colchicum autumnale

All parts of this plant are poisonous due to the presence of alkaloids, with particular reference to the corms and seeds. Found in meadows, mainly in the southern and western parts of Britain, and is cultivated. The flowers are six-petalled, light violet, pale purple in colour, and similar to the crocus, but grow to a height of 30 cm. (1 ft.), and appear in August, September and October. The broad pointed leaves appear in the spring, growing around the seed vessel from the previous year's flowers.

SANDWORT, IRISH
Arenaria ciliata

Contains poisonous saponins. A prostrate plant up to 8 cm. long. The white flowers appear from May onwards to August. Grows in Ireland.

SANDWORT, SMALL THYME-LEAVED
Arenaria leptoclados

Contains poisonous saponins. Grows up to 15 cm. (6 in.) high. The leaves resemble those of thyme, and the small flowers have five white petals. Found on sand in southern England.

SANDWORT, THYME-LEAVED
Arenaria Serpyllifolia
Doubtful—probably contains poisonous saponins. A very slender plant having white, five-petalled flowers which blossom in June, July and August. Found in very dry places, such as in old walling.

SCABIOUS, DEVIL'S-BIT
Succisa Pratensis
Doubtful. Grows up to half a metre tall. The bluish-purple flower-heads of minute blossom appear from June to September. Found on damp ground everywhere.

SCABIOUS, SMALL
Scabiosa Columbaria
Doubtful. Grows up to 60 cm. (2 ft.) tall with thin leaves, and the mauve flower heads appear in July and August. Found on chalk subsoils.

SIERRA LAUREL
Leucothoe
Contains the poisonous andromedotoxin.

SILVERWEED
Potentilla Anserina
Doubtful. The five-petalled solitary yellow flowers appear in July and August. The leaves are very smooth to the touch. Found everywhere.

SNAKEROOT, WHITE
Eupatorium
Very poisonous. Grows to one metre tall with broadly-pointed leaves in opposite pairs on the stem. The flower clusters at the head of each stem consist of minute white blossoms. Favours a moist, shaded position.

SNOWBERRY
Symphoricarpos Rivularis
Doubtful. This is a small shrub with minute pink flowers which blossom from June to August. The white berries are 12 mm. (½ in.) in diameter. Found in hedges as an escape.

SNOWDROP
Galanthus Nivalis

The bulbs are poisonous, and the flowers should be viewed with suspicion.

SNOW-ON-THE-MOUNTAIN
Euphorbia

The poisonous constituent has caused death among children. An ornamental pot plant.

SOAPWORT, CROW-SOAP, OR HEDGE PINK
Saponaria Officinalis

Contains poisonous saponins, the effect on water being similar to that of soap. Causes gastro-enteritis. Grows up to 60 cm. tall, with pale pink, pleasantly aromatic flowers in dense heads, which blossom in late August. The creeping root is fleshy. Found all over Britain, chiefly in urban areas.

SOLOMONS SEAL
Polygonatum Multiflorum

The berries are poisonous. Found in limestone-area woods. Grows to 60 cm. with the thin, bell-like, white flowers appearing in rows on the stems in May and June. The berries are black.

SORGHUM

Should be boiled for at the least $\frac{1}{4}$ hour, or fermentation will yield prussic acid. The red-grained variety was formerly used in beer brewing.

SORREL, COMMON, SOUR OR SOUR DOCK
Rumex Acetosa

Contains large amounts of oxalic acid, which the wine-maker would normally remove with precipitated chalk, when no harm should eventuate. Common in fields, wastes and gardens, being up to 60 cm. high, with leaves up to 13 cm. long, almost rectangular, green, terminating in a cluster of green flowers which finally turn reddish. Common on meadows and pastures having an acid soil.

SORREL, FRENCH
Rumex Scutatus

Contains large amounts of oxalic acid which can be

removed by the use of precipitated chalk. The leaves are up to 4 cm. long, and the same breadth. The flowers appear in June and July. Grows everywhere.

SORREL, SHEEPS OR SOUR GRASS
Rumex Acetosella

Probably best avoided in the amounts needed for wine-making, but a little can be included in salads, where the strong oxalic acid content is very noticeable. Grows up to 30 cm. high. The flower clusters growing up the stem are red in colour, appearing from May through to August. Common weed on pastureland, particularly on acid soils.

SORREL, WOOD
Oxalis Acetosella

The leaves are poisonous. Common in the shade of woods. The five-petalled white flowers with purple veins appear in April and May. The leaves are trefoil-type.

SOWBREAD
Cyclamen Hederifolium

Doubtful. The pinkish flowers with a touch of purple at their base appear in August and September. The stems are brownish, and the leaves shaped like those of rhubarb. Found in woodlands in southern England.

SPEARWORT, GREATER
Ranunculus Lingua

Poisonous. Grows up to 2 metres high. The five-petalled flowers are bright yellow in colour, and 5 cm. across, appearing from June to September. Occasionally found in wet places.

SPEARWORT, LESSER
Ranunculus Flammula

Poisonous. Grows up to 30 cm. in height. The flowers are 12 mm. across, a clear yellow colour, five-petalled, and appear from May through to September. Found in wet places throughout Britain.

SPINACH
Spinacia

The oxalic acid is best neutralised by precipitated chalk.

The flowers appear from June through to September. Found on chalky subsoil in south-western Britain.

SPURGE, WOOD
Euphorbia Amygdaloides
Poisonous. The flowers and leaves have more than a hint of yellow colouring. The flower stalks emerge from the head of the stem, and the flowers blossom from March to June. Grows everywhere in southern England.

SQUILL
Scilla Verna
Poisonous due to the glycoside content. The flowers are pale blue; a member of the bluebell family.

SQUIRREL CORN
Dicentra
Contains poisonous alkaloids.

STAR OF BETHLEHEM
Ornithogalum Umbellatum
Contains toxic alkaloids. The green leaves have a white rib and large white flowers which blossom from April to June.

STITCHWORT, LESSER
Stellaria Graminea
Doubtful. Grows to 1 metre tall on delicate stalks having bright green, pointed, thinly elliptical pairs of leaves. The white, star-like flowers appear from May through to October. Found everywhere in grassland.

ST. JOHN'S WORT, COMMON
Hypericum Perforatum
Could cause skin trouble, and is best avoided; particularly the flowers and leaves. This species of Hypericum is found everywhere in these islands, except perhaps in northern Scotland. It has a two-ridged stem which stands erect, up to 1 metre tall, carrying the smooth, stalkless leaves in opposite pairs; recognised by their translucent specks. The flowers have five clear yellow petals, and appear in July, August and September.

St. John's Wort, Hairy
Hypericum Hirsutum
Poisonous. Grows up to 60 cm. tall on an erect stem carrying pairs of pointed elliptical leaves without stalks, from which spring stalks of flower-clusters. The five-petalled, yellow, star-like flowers blossom in July and August. Found on the edges of woods and copses.

St. John's Wort, Imperforate
Hypericum Maculatum
Poisonous. Similar to the last-given species, but the flowers are a deeper yellow, and appear in July, August and September. Found on damp ground in Scotland.

St. John's Wort, Mountain
Hypericum Montanum
Poisonous. The leaves are elliptical in shape, pointed at the free end and joined direct to the stem. The pale yellow flower clusters appear at the head of the stem in July and August. Grows on chalk subsoils.

St. John's Wort, Slender, or Beautiful
Hypericum Pulchrum
Poisonous. Grows on a slender, erect stem, the stem leaves widely spaced, and the five-petalled, yellow flowers, reddish on the underside, blossom in June, July and August. Found everywhere on dry hillocks.

St. John's Wort, Marsh or Bog
Hypericum Elodes
Poisonous. The greyish-green stems and leaves are often floating. The yellow flower petals are upright, closed, and with a reddish base, appearing in July and August. The name is descriptive of the location of this plant.

St. John's Wort, Square-Stemmed
Hypericum Tetrapterum
Poisonous. The rectangular stalk carries pairs of oval-shaped leaves and the flower cluster at the head is of five-petalled yellow flowers, of which the buds are red, appearing in July and August. Found in wet places.

TOMATO
Lycopersicum Esculentum
The stems and leaves are poisonous.

TORMENTIL
Potentilla Erecta
Doubtful. The stalks and flower stems are slender, the leaves a muddy green, and the four-petalled yellow flowers appear from June through to September. Common in hilly districts on acid soils.

TRAVELLER'S JOY, OR OLD MAN'S BEARD
Clematis Vitalba
All parts poisonous. The fruits form a white, hairy, "old man's beard." The greenish-white flowers appear in June and July. Alkaline soils in southern England favour this plant, of which there are cultivated varieties.

TREFOIL, BIRD'S-FOOT
Lotus Corniculatus
Doubtful for winemaking and best avoided. A very common plant. There are five to eight flowers in a head, and they are bright yellow in colour, appearing from June to August.

TULIP
The bulbs are poisonous, and the flowers are best avoided when winemaking.

TUNG TREE
Aleuritas
Contains poisonous saponins in all parts. The nuts can be mistaken for Brazils, and one such can cause severe illness or death.

TURNIP
Best used in amounts which will ensure that the intake as wine will not be in excess of a normal meal.

VALERIAN
Valeriana Officinalis
Doubtful for winemaking. A thick-stemmed plant growing up to 1 metre tall, and carrying clusters of small, pink, bell-shaped flowers which blossom from June through to September. Grows in damp situations.

VETCH, COMMON
Vicia Sativa
Best avoided. Cultivated. Grows to 45 cm. high.
Purple flowers appear in pairs in May, June and July.

VETCHLING, HAIRY
Lathyrus Hirsutus
The seeds are dangerous. Grows up to 60 cm. tall. The
typical sweet-pea type flowers blossom in June and July.
Common on cultivated fields.

VETCHLING, YELLOW
Lathyrus Apacha
Poisonous in all parts. The yellow flowers blossom from
May through to August. Found on dry soil with a chalk
subsoil.

WATER DROPWORT, OR WATER HEMLOCK
Oenanthe Crocata
All parts of the plant contain a convulsant poison
which is not removed by drying and storage. Grows up to
$1\frac{1}{2}$ metres high with a hollow, grooved stalk. The leaves are
similar to those of celery. The flowers are white, in large
clusters at the top of the plant. Found in wet places everywhere except northern Scotland.

WATER DROPWORT, PARSLEY
Oenanthe Lachenalii
All parts poisonous, and the resinous poison is resistant
to drying and storage. The brilliant white flower clusters are
at the top of the stalks, and appear from June through to
September. Grows in salty marshland.

WAYFARING TREE
Viburnum Lantana
Doubtful for winemaking. The thick, yellowish-white
flower clusters appear in June and July. Grows in the hedgerows of southern England, particularly on chalky subsoils.

WHEAT
Should be boiled for at the least $\frac{1}{4}$ hour when used for
winemaking.

WILLOW HERB, GREAT HAIRY
Epilobium Hirsutum

So powerfully astringent that poisoning can result. Not to be confused with Rosebay or Fireweed, which has been used in brewing. Grows up to 1¼ metres tall on hairy stems, and with medium-oval pointed hairy leaves. The four-petalled pinkish-red flowers have white centre-pieces, and appear in July and August. Found by riversides, as distinct from the clearance sites favoured by Rosebay.

WISTERIA
Wisteria

The seeds and pods are poisonous.

WOAD
Isatis Tinctoria

To be avoided when winemaking. Grows to 1 metre in height. The hanging black seeds are easily recognised. Cultivated.

YELLOW RATTLE OR HAYRATTLE
Rhinanthus Minor

Poisonous. The stem is brownish, the narrow leaves toothed, and the yellow flowers emergent from a pale green base appear in May and June. Found on pastureland everywhere.

YEW, COMMON, AND IRISH
Taxus Baccata

The most poisonous of all trees native to this country, and all parts of the tree are poisonous, due to the presence of toxine. The branches are almost horizontal, but those of the Irish Yew are more erect. Grows up to 6 metres tall. Widely planted as an ornamental tree.

YEW, JAPANESE
Taxus

Contains poisonous alkaloids.